YOUR KNOWLEDGE HAS VALUE

- We will publish your bachelor's and master's thesis, essays and papers

- Your own eBook and book - sold worldwide in all relevant shops

- Earn money with each sale

Upload your text at www.GRIN.com and publish for free

Abhijeet Singh

Effect of stress on plant Proteome

GRIN Verlag

Bibliografische Information der Deutschen Nationalbibliothek:

Die Deutsche Bibliothek verzeichnet diese Publikation in der Deutschen National-bibliografie; detaillierte bibliografische Daten sind im Internet über http://dnb.d-nb.de/ abrufbar.

Imprint:

Copyright © 2013 GRIN Verlag GmbH
Druck und Bindung: Books on Demand GmbH, Norderstedt Germany
ISBN: 978-3-656-46996-4

This book at GRIN:

http://www.grin.com/en/e-book/230606/effect-of-stress-on-plant-proteome

CONTENTS

<u>ABBREVIATIONS</u>

GLP	Germin Like Protein
OxO	Oxalate Oxidase
SOD	Superoxide Dismutase
K_m	Michaelis Constant
V_{max}	Maximum Rate
NaCl	Sodium Chloride
pH	Parameter Hydrogen Ion
Etc	ET Cetera
EC	Electrical Conductivity
$NaHCo_3$	Sodium Bicarbonate
Na_2CO_3	Sodium Carbonate
Na_2SO_4	Sodium Sulphate
Å	Angstrom (10^{-10m})
SDS	Sodium Dodecyl Sulphate
mM	Milli Molar

<u>INTRODUCTION</u>

This review summarizes the Germin and Germin like proteins (GLPs), in relation to the crystal structure, amino acid sequence, biochemical properties, differential expression of proteins of subfamily under the salt stress. The interactive effects of the alkaline and salt stress on the Germin and Germin like protein will be discussed.

<u>SALT STRESS</u>

Plants, unlike animals, are sessile (Shanker A. 2011), and cannot move from one place to another in response to the unusual environment or any biotic and abiotic stress, they encounters. To protect themselves they need to change in their internal environment to face the situation and to survive in it. There are many stresses in nature which occurs time to time at various extents, for example- Biotic stress – fungi, bacteria, viruses, nematodes, insects, animals etc. and Abiotic stress – drought, flood, high & low temperature, pH, salinity , sodicity, alkalinity, acidity, ion deficiencies and toxicities. Plants respond to these stresses by changes and modification in biophysical, biochemical, physiological, morphological and developmental processes (Caliskan. M; 2011). Salt stress is one of the major abiotic stresses faced by the plants. Soil salinity and alkalinity seriously affect about 932 million hectares of land globally, reducing productivity in about 100 million hectares in Asia (Rao et al. 2008, Wang et al. 2011).

Soil salinity in agriculture soils refers to the presence of high concentration of soluble salts in the soil moisture of the root zone. Soils are considered saline or salt affected when the electrical conductivity of water extracted (EC) from water-saturated soil samples from the root zone exceeds 4 dSm^{-1} (Richards, 1954). These high concentrations of soluble salts through their high osmotic pressures affect plant growth by restricting the uptake of water by the roots. Salinity can also affect plant growth because the high concentration of salts in the soil solution interferes with balanced absorption of essential nutritional ions by plants (Tester and Devenport, 2003). Reports have clearly demonstrated alkaline salts ($NaHCO_3$ and Na_2CO_3) are more destructive to plants than neutral salts (NaCl and Na_2SO_4) (Shi and Sheng 2005, Shi and Wang 2005, Yang et al. 2008 a,b,c ; Wang et al. 2011). Salt stress in a soil generally involves osmotic stress and ion-induced injury (Munns 2002), there is an added high pH effect with alkali stress. A high-pH environment surrounding the roots can cause metal ions and phosphorus to precipitate, with loss of the normal physiological functions of the roots and destruction of the root cell structure (Li et al. 2009). Alkali stress can inhibit absorption of inorganic anions such as Cl^-, NO_3^- and $H_2PO_4^-$, greatly affect the selective absorption of K^+-Na^+, and break the ionic balance (Yang et al. 2007, 2008b, 2009). Thus, plants in alkaline soil must cope with physiological drought and ion toxicity, and also maintain intracellular ion balance and regulate pH outside the roots.(Wang et al. 2011)

GERMIN AND GERMIN LIKE PROTEIN

Germin and Germin-like proteins (GLPs) constitute a large and highly diverse family of developmentally regulated proteins showing a wide range of distribution ubiquitously from Myxomycetes to flowering plants. Germin was initially identified as a specific marker for the start of germination in wheat embryos, from which function it was given the name "Germin" (Thompson and Lane, 1980). Germin genes and their proteins were first detected in Germinating cereals (Grzelczak et al., 1985), but subsequently, Germin-like proteins were also identified in dicotyledonous angiosperms (Michalowski and Bohnerd, 1992), gymnosperms (Domon et al., 1995) and mosses (Yamahara et al., 1999).

CRYSTAL STRUCTURE OF GERMIN

Woo et al. (1998, 2000) determined the structure of Germin at 1.6 Å resolution, and showed that the mature protein comprises six β-jellyroll monomers locked into a homohexamer (a trimer of dimers){fig. 1}.This structure accounts for its remarkable stability to various denaturing agents; all germins share unusual resistance to broad specificity proteases and to dissociation by various agents such as heat, SDS and extreme pH (Lane et al., 1993; Lane, 1994; Wei et al., 1998; Carter and Thornburg, 2000; Membré et al., 2000).

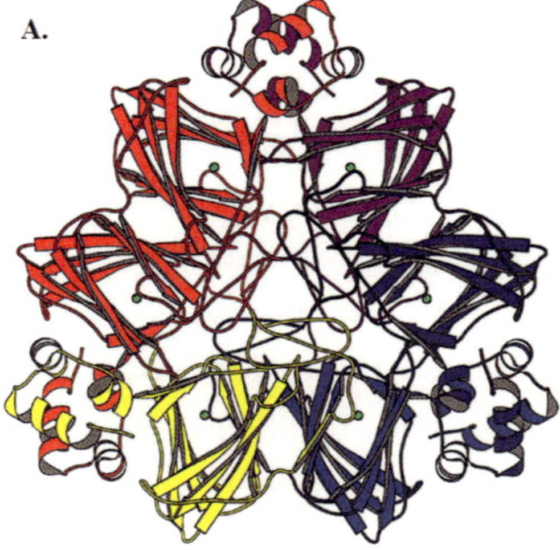

Fig.1 - A. The germin homohexamer with six manganese ions indicated (green spheres). Note the x-helical clasps that bind the dimers into the trimer-of-dimers and the tight packing around the three-fold axis. The pro-motif of the sequence is with 2 sheets, 2 beta hairpins, 4 bulges, 10 strands, 6 helices, 3 helix-helix interace, 17 beta turns, 1 gamma turn, 1 disulphide bond. (http://www.ebi.ac.uk/pdbsum/1fi2)

The amino acid sequence of the Germin is 201 AA long which is described with the sequence annotation by sequence in fig. 2

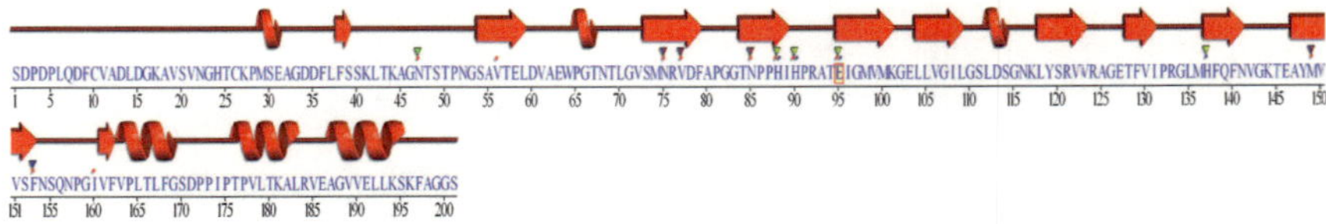

Fig. 2 – The amino acid sequence is 201 residue, the spiral represents helix and arrow strand, disulfide bond occurs between AA 10 & 26, AA 95 has catalytic residue,triangle represents active sites, dot (·) represents residue contact to metal (blue) & to ligand (red), 2 beta hairpin occurs at AA 60-72 & 110-117. (http://www.ebi.ac.uk/pdbsum)

ISOFORMS OF GERMIN

There are isoforms in existence of Germin and three isoforms have been defined. These are G, G' and ΨG (pseudoGermin). The G and G' oligomers are water soluble, resistant to digestion by pepsin and to dissociation in aqueous SDS solutions at room temperature (Lane, B. G. and Tumaitis-Kennedy, T. D., 1981; Grzelczak, Z. F. and Lane, B. G., 1983). Amino acid composition and sequencing indicate that G and G' have the same apoprotein and differ only in that G has two further N-acetylglucosamine units attached to the basic core of its N-glycans (Lane et al 1990). Both forms of Germin were defined as glycoproteins. The carbohydrate content of Germin is about 10% by weight (Lane et al 1990). In addition to the Germination related Germin isoforms (G and G'), and antigenically related homo-tetrameric form of Germin (pseudoGermin-ΨG) has also been detected, particularly in the cell walls of immature wheat embryos at a time when the maximum cellular enlargement associated with embryogenesis when maturation is occurring (20-25dpa). Unlike Germin, pseudoGermin is thermostable; the oligomer remains undissociated even when boiled in the presence of SDS, so long as non-reducing conditions (Lane et al 1992).

<u>BIOCHEMICAL PROPERTIES</u>

These proteins have been cited in various plant organs like roots, leaves, nectar glands and seeds etc. This protein is localized in the cell wall and is a water soluble glycoprotein with oxalate oxidase (OxO) activity that converts oxalate to hydrogen-per-oxide (Lane *et al.*, 1993). The subgroups of GLPs have different enzyme functions that include the two hydrogen peroxide-generating enzymes, oxalate oxidase (OxO) and superoxide dismutase (SOD). Cereal Germin protein is a homo-pentameric apoplastic glycoprotein whose synthesis is associated with the onset of growth in Germinating wheat embryos (Lane, 1991). Germins exists in the oligomeric forms and are highly resistant to proteases and heat treatment (Lane, 1994; Dunwell *et al.*, 2000). Proteins with sequence similarity to Germins have been identified in diverse plant groups and are known as Germin-like proteins (GLPs). GLPs are also members of cupin superfamily of proteins like Germins due to the presence of cupin motif in their amino acid sequence (Dunwell *et al.*, 2004). Majority of Germins identified so far exhibit oxalate oxidase activity which is associated with many stress responses and developmental processes (Wei *et al.*, 1998; Patnaik and Khurana, 2001; Liang *et al.*, 2001). Germin-like proteins also seem to be involved in many stress related processes, but their definite role in these processes remains obscure. (Yasmin T. 2009). Germins and GLPs are thought to play a significant role during zygotic and somatic embryogenesis (Caliskan et al. 2004; Neutelings et al. 1998; Mathieu et al. 2003), salt stress (Berna et al. 1999), drought stress (Ke et al. 2009), pathogen elicitation (Jøhnk et al. 2005; Godfrey et al. 2007; Manosalva et al. 2009), and heavy metal stress (Zhou et al. 2009), etc.

GERMIN AND GERMIN LIKE PROTEIN RELATED GENES

A large number of GLP genes have been discovered as expressed sequence tags (ESTs) or by genome sequencing in higher plants like Arabidopsis (Carter et al., 1998; Membré et al., 1997), barley (Wu et al., 2000, Druka et al., 2002), and rice (Membré and Bernier, 1998). Bernier and Berna (2001) reported that the number of different sequences related to Germin is now close to a hundred, also the number of GLP genes in each species of higher plant exceeds 30 (Cannon et al., 2004) and most are present within the genome as multiple copies at a small number of loci (Kazusa et al.,2000; Manosalva, 2006).

The identification of genes whose expression enables plants to adapt to or tolerate to salt stress is essential for breeding programs, but little is known about the genetic mechanisms for salt tolerance. One approach in clarifying the molecular mechanisms involved in salt stress is to identify the genes whose levels change as a result of salt stress. Hurkman et al. (1989) reported that in barley, gene regulation is altered by salt stress and the levels of translatable mRNAs change with salt treatment. Among the salt stress responsive gene products, Germin and Germin-like proteins (GLP) were identified (Caliskan, 1997; Hurkman et al., 1989).

SALT INDUCED GENE EXPRESSION OF GERMIN

Isolation and examination of the two Germin genomic clones (Rahman et al., 1988) and the determination of the predicted amino acid sequences have revealed high homology with spherulin 1a/1b proteins of the slime mould *Physarum polycephalum* (Lane et al., 1991). The synthesis of these proteins occurs during spherulation: a transition leading to developmental arrest imposed by environmental conditions such as osmotic stress and starvation (Bernier *et al.*, 1987), this similarity led to the suggestion that another possible function for Germin might be related to the changing osmotic properties of cells. In support of this notion, synthesis of Germin-like proteins was discovered to be altered upon salt stress in barley (Hurkman et al., 1991) and in the halophytic "Ice plant", *Mesembryanthemum crystallinum*, (Michalowski & Bohnert, 1992). In salt-stressed barley, wheat GLPs were resistant to protease and were glycosylated and heat stable. They were detected in barley roots (but not in tips) and coleoptiles but not leaf of 6 day seedlings. Their synthesis increased in roots upon salt stress, but decreased in coleoptiles. Thus, these different studies implied that Germin might represent a family of proteins of which individual members may have different biochemical functions related to changes of the osmotic properties of the cell.

Messenger RNA analysis in different plant suggests that germin and GLPs have different special distribution of the mRNA expression (Hurkman & Tanaka, 1996a). The vascular transition region was reported to contain the highest levels of germin mRNA in wheat, whereas roots displayed the highest germin expression levels in barley seedling. Cereal germin proteins have strong oxalate oxidase activity (Lane et al., 1993), an activity that produces one mole of H_2O_2 and two moles of CO_2 from degradation of oxalic acid. It is reported that H_2O_2 might act as a signaling molecule at low concentration (Luthell, 1993) or a component of cell wall modifications at high concentrations (Showalter, 1993). Another germin-like protein isolated from the cells of a moss, *Barbula unguiculata*, was shown to have Manganese superoxide dismutase activity (Yamahara et al., 1999). Germin genes and proteins have been shown to be associated with various aspects of plant development (Caliskan, 2000; Lane, 2002) such as defense system (Berna and Bernier, 1999; Donaldson et al. 2001), embryonic development (Caliskan, 2001), photoperiodic oscillations (Ono et al. 1996), and hormonal stimuli (Berna and Bernier, 1997).

Germin and germin-like proteins are suggested to be salt-responsive gene products and their response to salt stress seems to be various. For example, accumulation of germin mRNA is up-regulated during the growth of germinating barley seedlings in the presence of NaCl (Hurkman & Tanaka, 1996a). Contrary in ice plant (*Mesembryanthemum crystallinum*), it is reported that the synthesis of GLPs declined after salt stress (Michalowski and Bohnert, 1992). However, it was reported that in wheat seedlings germin synthesis was remained unchanged in the presence of NaCl (Berna & Bernier, 1999; Caliskan, 2009).

The model plant Arabidopsis has also been used in such studies. For example, in a proteomic analysis, it was shown that two GLPs, GLP9 (At4g14630) and OxO-like protein (At5g38940), increased in abundance in Arabidopsis roots subjected to NaCl treatment (Jiang et al., 2007). Similar results have also been reported from an investigation of the gene expression profile of third leaves of rice (cv. Nipponbare) seedlings subjected to salt stress (130 mM NaCl) (Kim et al., 2007).

In situ RNA hybridization is one of the most powerful techniques developed for localizing the expression site of particular gene products at the cell, tissue and organ levels. This method is particularly useful in understanding the function of specific gene products in particular tissues and the relation between tissue function and its localization in the whole structure of an organ (Ranjhan et al., 1992). This technique was employed for analyzing the possible germin functions and it was shown that germin mRNAs synthesized in the cells of coleorrhiza in wheat seedlings and it was considered that the enzymatic activity of germin, oxalate oxidase, might play an important role in metabolic regulation, particularly in cell wall modification during germination and seedling development (Caliskan & Cuming, 1998). It is well known that stress factors alter the synthesis of gene expression. Indeed, upon salt stress the localization pattern of germin gene expression was changed (Caliskan, 2009). It is shown that although the water grown embryos and salt stress grown embryos accumulate the similar amount of germin mRNAs, the synthesis site of germin mRNAs are completely different from each other.

Another comprehensive study (Berna and Bernier, 1997; 1999) utilizing a promoter-glucuronidase (GUS) fusion and showing induction of the wheat germin promoter in transgenic tobacco treated with salt, heavy metals, aluminium and plant growth regulators, specifically auxin and gibberellin. Expression of these genes, together with OxO enzyme activity and the H_2O_2 content of stressed seedlings, were induced by stress during germination conditions in this good emerging hybrid and were not induced in a variety that emerged poorly. It was postulated that a block in oxalate metabolism contributed to lower germination under stress in the low emerging variety.

In the study with *Nicotiana tabacum* plants as a model (Dani et al. 2005) investigated changes in the soluble apoplastic composition induced in response to salt stress. Apoplastic fluid was extracted using a vacuum infiltration procedure from leaves of control plants and plants exposed to salt stress. Two-dimension electrophoretic analyses and mass spectrometry revealed the identity of 20 polypeptides whose abundance changed in response to salt stress. While the levels of some proteins were reduced by salt-treatment, an enhanced accumulation of protein species known to be induced by biotic and abiotic stresses was observed. In particular, two chitinases and a GLP increased significantly. Similarly, expression of a germin protein was shown to be altered in tomato seedlings in response to salt stress (Amini et al., 2007).

In the biochemical study of the OxO and GLP expression in vivo, Singh et al., 2006, reported the effect of NaCl on the activity of the enzyme in vitro. The effect of NaCl stress on molecular and biochemical properties of OxO was studied in the seedling leaves of a grain sorghum hybrid. There was no effect on molecular weight and number of subunits of the enzyme but it did show some important changes in its kinetic parameters such as Km for oxalate and V_{max}. Optimum pH (5.8), activation energy (5.08 kcal mole^{-1}), time of incubation (6 min) and Km for oxalate (1.21×10^{-4}M) were increased, while V-max (0.18 mmole min^{-1}) decreased and no change in optimum temperature was observed. This showed that substrate affinity and maximum activity of the enzyme were adversely affected. The specific activity of OxO was also increased in seedlings grown in a NaCl containing medium compared to normal, which reveals the increased de novo synthesis of the enzyme to sustain oxalate degradation.

CONCLUSION

Although, Germin and Germin-like proteins have been studied extensively since the 1980s, their biological importance and functions remain confusing. Furthermore, germin, germin like proteins and oxalate oxidase have been found in a broad range of various plant species under different circumstances, and related with different aspects of plant development. For example, germin-like protein and oxalate oxidase enzyme activity have been identified in plants in relation to salt stress, pathogen infection, photoperiodic oscillations, germination and embryogenesis. It has also reported that germin and germin-like proteins and oxalate oxidase are responsive to various plant growth regulators such as auxin and abscisic acid. Germin-like oxalate oxidase therefore seems to fulfill some crucial functions in plants. In addition to these properties, recently it was suggested that germins and various other proteins which are involved in early plant development might belong to an evolutionarily ancient superfamily: cupin superfamily (Dunwell et al., 2000; 2008).

<u>REFERENCES</u>

1. Amini, F., Ehsanpour, A. A., Hoang, Q. T., and Shin, J. S. 2007. Protein pattern changes in tomato under in vitro salt stress. Russian J. Plant Physiol. 54: 464-471.

2. Bernier, F. and A. Berna. 2001. Germins and germin-like proteins: plant do-all proteins. But what do they do exactly? *Plant Physiol Biochem.,* 39: 545-554

3. Bohnert, H. J., Ostrem, J. A., Cushman, J. C., Michalowski, C. B., Rickers, J., Meyer, G., DeRocher, E. J., Vernon, D. M., Krueger, M., Vazquez-Moreno, L., Velten, J., Höfner, R., and Schmitt, J. M. 1988. Mesembryanthemum crystallinum, a higher plant model for the study of environmentally induced changes in gene expression. Plant Mol. Biol. Rep. 6: 10-28.

4. Caliskan, M. (2000). Germin, an oxalate oxidase, has a function in many aspects of plant life.*Tr. J. Biol.* 24, 717-724

5. Caliskan, M. (2009). Salt stress causes a shift in the localization pattern of germin gene expression. *Gen. Mol. Res.* 8, 1250-1256

6. Caliskan, M., Bashiardes, S., Ozcan, B. & Cuming, A.C. (2003). Isolation and localization of new germination-related sequences from wheat embryos. *J. Biochem. Mol. Bio.* 36(6), 580-585

7. Caliskan, M., Turet, M. & Cuming, A.C. (2004). Formation of wheat (Triticum aestivum L.) embryogenic callus involves peroxide-generating germin-like oxalate oxidase.*Planta* 219, 132-140

8. Caliskan, M., Turet-Sayin, M., Turan, C. & Cuming, A. (2004). Identification of germin isoforms in wheat callus. *Cereal Res. Com.* 32, 355-361

9. Guo. Y. And Song Y., Differential proteomic analysis of apoplastic proteins during initial phase of salt stress in rice, Plant Signaling & Behavior 4:2, 121-122.

10. H. Wang, z. Wu, y. Chen, c. Yang, d. Shi, effects of salt and alkali stresses on growth and ion balance in rice (*oryza sativa* l.), plant soil environ., *57*, 2011 (6): 286–294

11. Harkamal W., Wilson C., Wahid A. Condamine P, Cui X., Close X. J., Expression analysis of barley (Hordeum vulgare L.) During salinity stress, Funct Integr Genomics, DOI 10.1007/s10142-005-0013-0

12. Hurkman W J and Tanaka C K 1987 The effects of salt on the pattern of protein synthesis in barley roots. Plant Physiol. 83, 517-524.

13. Hurkman W J and Tanaka C K 1988 Polypeptide changes induced by salt stress, water deficit, and osmotic stress in barley roots: A comparison using two-dimensional gel electrophoresis. Electrophoresis 9, 781-787.

14. Hurkman W J, Fornari C S and Tanaka C K 1989 A comparison of the effect of salt on polypeptides and translatable mRNAs in roots of a salt-tolerant and a saltsensitive cultivar of barley. Plant Physiol. 90, 1444-1456.

15. Hurkman W J, Tao H P and Tanaka C K 1991 Germin-like polypeptides increase in barley roots during salt stress.Plant Physiol. 97, 366-374.

16. Jiang, Y., Yang, B., Harris, N. S., and Deyholos, M. K. 2007. Comparative proteomic analysis of NaCl stress-responsive proteins in Arabidopsis roots. J. Exp. Bot. 58: 3591-3607.

17. Lane B G, Gzelczak Z, Kennedy T, Hew C and Joshi S 1987, Preparation and analysis of mass amounts of germin: Demonstration that the protein which signals the onset of growth in germinating wheat is a glycoprotein. Biochem. Cell. Biol. 65, 354

18. Lane, B. G., Dunwell, J. M., Ray, J. A., Schmitt, M. R., and Cuming, A. C., Germin, a protein marker of early plant development, is an oxalate oxidase. J. Biol. Chem., 268: 12239-12242, 1993.

19. LI Hui-yu • JIANG Jing • WANG Shan • LIU Fei-fei, Expression analysis of *thglp*, a new germin-like protein gene, in *Tamarix hispida,* Journal of Forestry Research (2010) 21(3): 323–330

20. Naqvi, S. M. S. 1993. Identification characterization and sub cellular localization of salt induced

21. proteins in rice (Oryza sativa). Ph.D. Thesis, Middle East Technical University, Ankara, Turkey.

22. Ramagopal, S. 1987. Differential mRNA transcription during salinity stress in barley. Proc. Nat. Acad. Sci. USA 84: 94-98.

23. Shanker A., Abiotic Stress Response in Plants - Physiological, Biochemical and Genetic Perspectives, ISBN 978-953-307-672-0

24. Tayyaba y., tariq m., m. Zeeshan hyder, samina a. And s. M. Saqlan naqvi, cloning, sequencing and *in silico* analysis of germin-like protein gene 1 promoter from *oryza sativa* l. Ssp. Indica, *pak. J. Bot.*, 40(4): 1627-1634, 2008.

25. Wang, M.L., Chen, X., Holbrook, C., Culbreath, A., Liang, X., Brenneman, T. & Guo, B. (2011). Identification and characterization of a multigene family encoding germinlike proteins in cultivated peanut (Arachis hypogaea L.). *Plant Mol. Biol. Rep.* 29, 389–403

26. William j. Hurkman , effect of salt stress on plant gene expression: a review, *plant and soil* 146: 145-151, 1992.

27. Woo, E. J., Dunwell, J. M., Goodenough, P. W. and Pickersgill, R. W. 1998. Barley oxalate oxidase is a hexameric protein related to seed storage proteins: evidence from X-ray crystallography. FEBS Lett. 437: 87-90.

28. Woo, E. J., Dunwell, J. M., Goodenough, P. W., Marvier, A. C., and Pickersgill, R. W. 2000. Germin is a manganese containing homohexamer with oxalate oxidase and superoxide dismutase activities. Nat. Struct. Biol. 7: 1036-1040.

29. Zhang, Z., Collinge, D. B. and Thordal-Christensen, H., Germin-like oxalate oxidase, a H2O2-producing enzyme, accumulates in barley attacked by the powdery mildew fungus. The Plant Journal, 8 (1): 139-145, 1995.

30. Zouari, N., Saad, R. B., Legavre, T., Azaza, J., Sabau, X., Jaoua, M., Masmoudi, K., and Hassairi, A. 2007. Identification and sequencing of ESTs from the halophyte grass Aeluropus litoralis. Gene 404: 61-69.